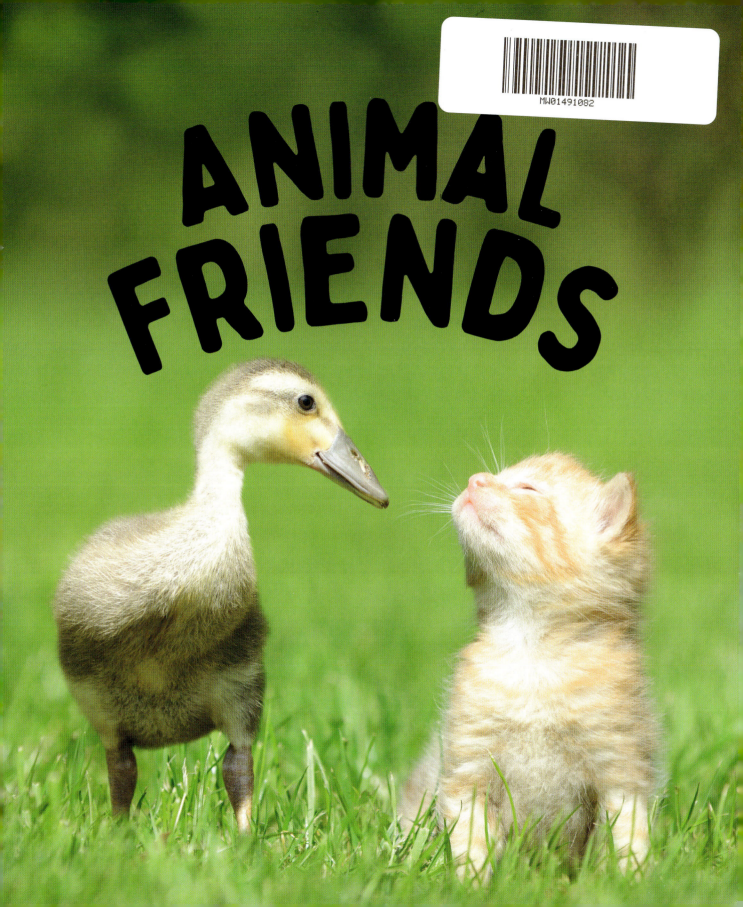

ANIMAL FRIENDS

Inspiring | Educating | Creating | Entertaining

Brimming with creative inspiration, how-to projects, and useful information to enrich your everyday life, quarto.com is a favorite destination for those pursuing their interests and passions.

© 2022 Quarto Publishing Group USA Inc.

First published in 2022 by QEB Publishing, an imprint of The Quarto Group.
100 Cummings Center,
Suite 265D Beverly,
MA 01915, USA.
T (978) 282-9590 F (978) 283-2742
www.quarto.com

Author: Camilla De La Bedoyere
Associate Publisher: Holly Willsher
Designers: Susi Martin & Kevin Knight
Editorial Assistant: Alice Hobbs
Picture Researcher: Sarah Bell

A CIP record for this book is available from the Library of Congress.

ISBN: 978-0-7112-8042-7

9 8 7 6 5 4 3 2 1

Manufactured in Guangdong, China TT082022

FSC
www.fsc.org

MIX
Paper from responsible sources
FSC® C016973

CONTENTS

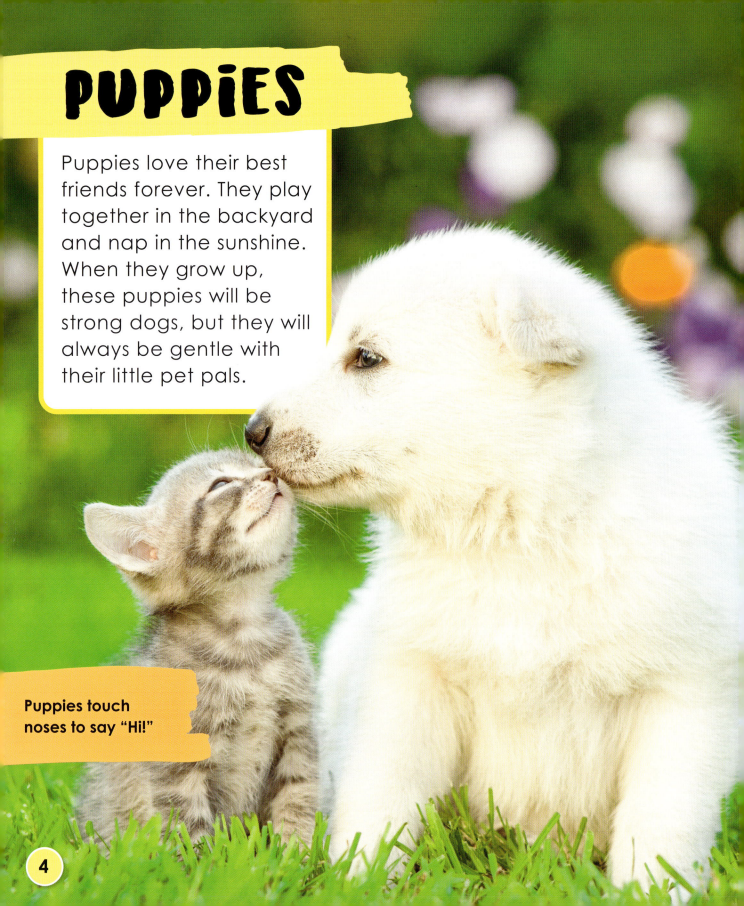

PUPPIES

Puppies love their best friends forever. They play together in the backyard and nap in the sunshine. When they grow up, these puppies will be strong dogs, but they will always be gentle with their little pet pals.

Puppies touch noses to say "Hi!"

You're my
buddy, bunny!

Chimps live with their families and friends and they take care of each other with lots of kisses and cuddles.

CHIMPANZEES

Cheeky chimps like playing and they love tickles! When they are tired, baby chimps cuddle with their mom, but sometimes any furry friend will do.

ELEPHANTS

I love you!

Baby elephants are best friends with their brothers, sisters, and cousins. An elephant's trunk is a long, bendy nose but it comes in handy for lots of things. Elephants hold each other's trunks, just like we hold hands!

Elephant best friends use their trunks to stroke and hug each other.

GROUND SQUIRRELS

As soon as the sun comes up, ground squirrels get busy. They scamper out of their burrows to say "Hi!" After a quick nuzzle and a hug, the friends like to play tag. They jump and tumble over each other, flick their tails, and then it's time to find some food.

Gotcha!

Ground squirrels are chatty animals. They snarl, grunt, whistle, and squeal.

Did you see that?

LEMURS

Like all best friends, lemurs sometimes fight—but most of the time they love to cuddle and have fun. Ring-tailed lemurs have long, striped tails that they use to wrap around themselves or their friends. Lemurs live on the island of Madagascar where they spend a lot of time in trees.

Lemurs live in big family groups called troops. They take care of each other.

KiTTENS

Kittens are curious baby cats. They like to see, sniff, and even lick their new friends! Their best buddies are soft and warm—kittens don't like to get chilly, so they love to snuggle up with a furry, or feathery, friend. After a catnap, a kitten turns into a bundle of energy and wants to play.

I'm coming to get you!

Kittens like to play, but they must learn how to be gentle with their friends.

15

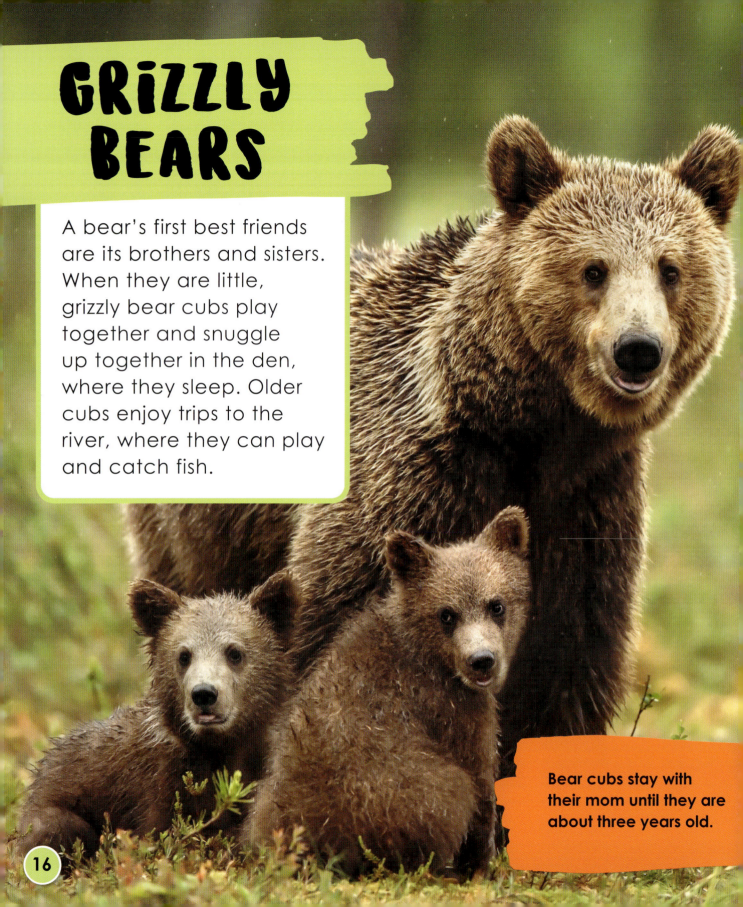

GRIZZLY BEARS

A bear's first best friends are its brothers and sisters. When they are little, grizzly bear cubs play together and snuggle up together in the den, where they sleep. Older cubs enjoy trips to the river, where they can play and catch fish.

Bear cubs stay with their mom until they are about three years old.

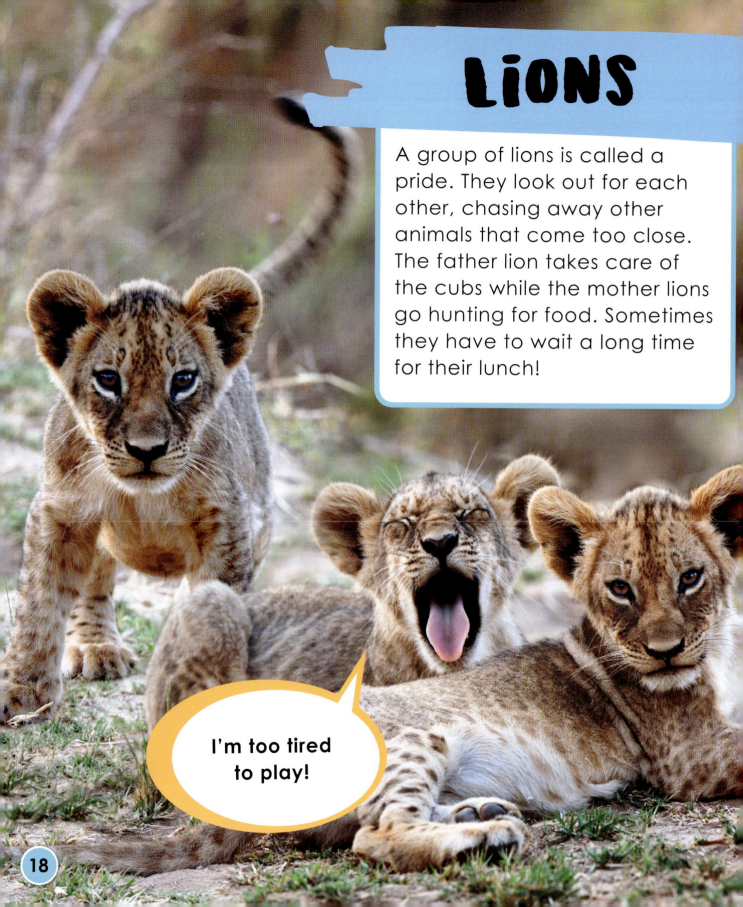

LiONS

A group of lions is called a pride. They look out for each other, chasing away other animals that come too close. The father lion takes care of the cubs while the mother lions go hunting for food. Sometimes they have to wait a long time for their lunch!

I'm too tired to play!

Lion cubs are playful,
but they enjoy a rest in
the sun, or hiding in a tree.

SNOW MONKEYS

When the first fluffy flakes of snow fall from the sky, these monkeys are happy. It may be cold, but they don't mind because they can cuddle their furry friends, or jump into the hot springs nearby. The baby monkeys have fun in the snow, making snowballs!

Snow monkeys live in Japan, in places where hot water bubbles up from the ground.

Right, you asked for it!

TIGERS

Young tigers are born in a den. They are blind and tiny when they are born—they weigh less than a pet cat! When a tiger cub is about four months old it's ready for playtime with its siblings. Tiger cubs pounce on their pals, they wrestle with them, and even pretend to bite them.

If playtime gets too wild the mom gently picks the cubs up and takes them back to the den.

Be gentle with me!

SEA OTTERS

Holding hands is a good way to show your best friend that you love them. Sea otters hold paws to keep their best friends from floating away! These soft, furry animals live in the ocean. Moms even hold their babies on their tummies while they swim.

Hold me tight!

Baby sea otters can't swim until they are about four weeks old.

25

PiGS

Oink, love you!

Pigs are clever, playful animals. They are ready to make friends just a few hours after they are born. Piglets make friends with their brothers and sisters at first, but they love to meet all sorts of farm animals! When a pig is happy, its curly tail points up and it chats with its pals using sounds such as coos, oinks, and grunts.

A mother pig is called a sow. She can have more than ten piglets at a time to take care of!

GOLDEN LION TAMARINS

These little monkeys love cuddling! Golden lion tamarins live in hot, rainy forests in South America. Each mother has twins, and she lets the dad take care of her babies. He carries them on his back and teaches them how to find food to eat. The twins play together, but they also clean each other's fur.

I want a piggyback ride!

A baby tamarin's fur is soft and silky. It can be light gold, dark gold or red-gold, and very shiny.

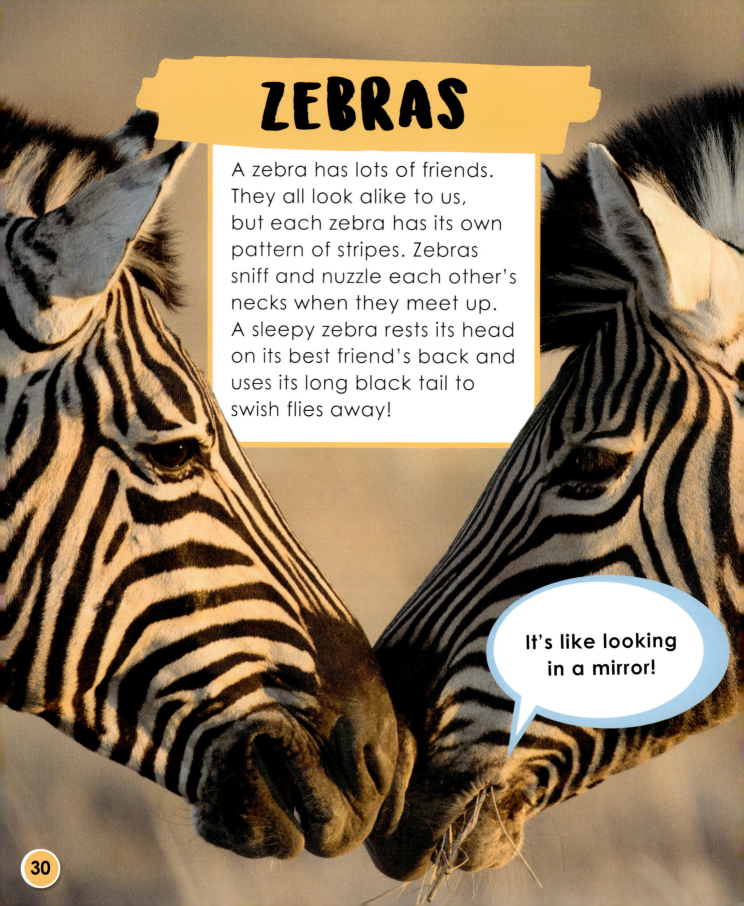

ZEBRAS

A zebra has lots of friends. They all look alike to us, but each zebra has its own pattern of stripes. Zebras sniff and nuzzle each other's necks when they meet up. A sleepy zebra rests its head on its best friend's back and uses its long black tail to swish flies away!

It's like looking in a mirror!

These stripy horses live in big groups called herds. Zebras live in Africa.

DUCKS

As soon as a duckling hatches from its egg, it's ready to make friends. In the first few hours of life, a duckling looks around for its mom, or any other animal. It will then want to stay with the animal it's found, no matter what kind it is!

Ducks are happy to follow their best friends everywhere.

Follow me!

GRAY WOLVES

Everyone knows that a dog makes a wonderful best friend. Gray wolves are very friendly wild dogs. They live in big families that hunt and play together in woods and forests. If a wolf can't find its friend, it howls loudly and then greets its lost playmate with a soft whimper or squeak to say "I found you!"

A group of wolves is called a pack. Young wolves are called pups or cubs.

I found you!

RABBITS

Rabbits are some of the world's friendliest animals. They can even make friends with cats and dogs, if they meet them when they are very young. Rabbits use their ears and feet to talk to their friends. If a rabbit's ears point up and forward, it is saying it's happy.

Let's chat!

When a rabbit thumps its legs, it's telling its friends that it is feeling angry.

MEERKATS

Young meerkats learn how to be good friends by watching older meerkats. They learn how to play and how to help each other find food. As the sun sets, the meerkat pups go back to their burrows, where they snuggle up for a rest.

We're watching you!

Meerkats have strong feet and claws. They use them to dig burrows.

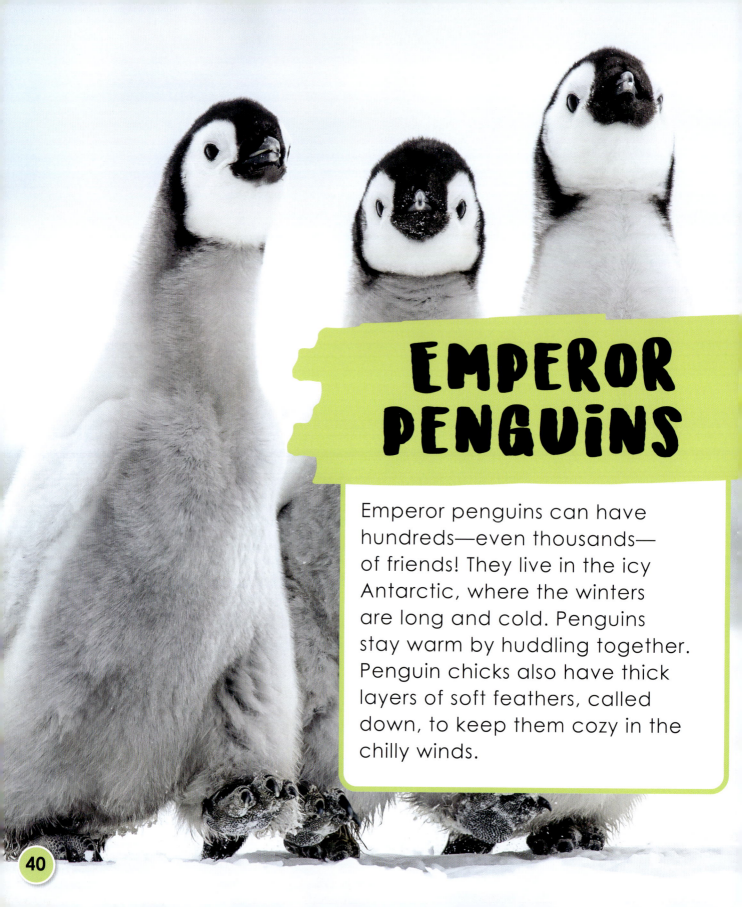

EMPEROR PENGUINS

Emperor penguins can have hundreds—even thousands—of friends! They live in the icy Antarctic, where the winters are long and cold. Penguins stay warm by huddling together. Penguin chicks also have thick layers of soft feathers, called down, to keep them cozy in the chilly winds.

Stay with me!

The chicks can't swim until they lose their fluffy gray feathers and grow new black-and-white ones.

You smell good!

PONIES

Ponies are small horses that are good at making friends. They love to live with other horses or ponies. Ponies run and jump when they play, waving their tail like a flag. They greet their friends by sniffing their noses and some ponies are happy to give their human pals a ride!

Ponies that live in cold places have thick fur and long manes.

43

WILD BOARS

Wild boar piglets are small animals and very friendly. At first, they play with their brothers and sisters but they get along with lots of other animals too. The piglets have stripy coats that help them hide in their woodland homes.

Let's play!

When they grow bigger, the piglets will lose their stripes.

ALPACAS

Alpacas are super-friendly, most of the time! They love to spend time with each other, but they also make friends with goats and sheep. If they get scared or angry, alpacas spit and kick—but they are quick to make friends again! They even hum sweetly to each other!

A mother alpaca has just one baby at a time. It is called a cria.

Let's hum a tune!

POLAR BEARS

Polar bear cubs are best buddies. They are born in a snowy den in the middle of winter. They stay in the den until spring comes, when they can start to explore. Cubs love to roll around in the snow together, and sometimes they can persuade their mom to join in too!

Thick fur helps keep polar bears warm in the Arctic, where they live.

I want a bear hug!

CHINCHILLAS

Big family groups of chinchillas live together in holes, high up on chilly mountains. They snuggle up close to stay warm during the day, but at night they scamper and jump around the rocks, eating grass and seeds by the moonlight.

A chinchilla is a fun, fluffy friend! Its fur is soft, thick, and very warm.

Furry friends
are fun!

Look at us, Mom!

SQUIRREL MONKEYS

Treetops make a perfect playground for little squirrel monkeys. They purr, cackle, and twitter at their friends as they snuggle or leap through the branches. Squirrel monkeys live in rain forests and they are always on the move, looking for tasty fruit to eat.

It's time for a ride! The baby monkeys cling on to their mom's back.

GIANT PANDAS

Giant panda bears are sleepy bears that live in bamboo forests in China. When the sun shines, the cubs know it's time to wake up and play! They love to do somersaults and roll around together. Panda bear cubs love to climb trees—just like we do!

Let's squeeze!

Playtime turns playful cubs into sleepy bears!

COYOTES

Coyotes are wild dogs that love a game of rough-and-tumble. When coyote cubs are just a few days old they start to play fight. To show their friends that a game is about to start, the cub bends its legs and lowers its head to the ground. This is a coyote's way of saying "Do you want to play with me?"

One is fun, two is best!

Playtime is the perfect way for cubs to learn how to hunt.

KOALAS

A baby koala's best friend is its mom! They love to be close and when they are very little, the babies stay in her soft pouch. When the baby is too big to snuggle in the pouch, the mom carries it on her back. Koalas spend most of their time in the treetops, sleeping or munching leaves.

I love my mom!

Koalas can sleep for up to 18 hours a day!

SEALS

These fun-loving seals will be best friends forever! Seal families live in the sea, where they find plenty of fish to eat. Young seals are called pups and they are very playful. When they are older, the pups will dive and dart through the waves, chasing each other.

Seals swim in the sea and sleep on land. But they can play anywhere!

Sleep now, play later!

Let's go faster!

HORSES

A friendly horse makes a sweet noise, called "nickering." It keeps its lips together while making a soft sound, and often touches its nose to the nose of its friend. This means "I am so happy to see you." Best friends also use their teeth to gently scratch and rub each other's backs.

Horses like being with farm animals, as well as cats and dogs.

PICTURE ACKNOWLEDGMENTS

Alamy
Alan Tunnicliffe 13b, Ariadne Van Zandbergen 8, BIOSPHOTO 49, Cavan Images 49t, Cultura Creative RF 15t, Danita Delimont 21, Eric Gevaert 29b, Evgeniy Baranov 22b, F1online digitale Bildagentur GmbH 45b, Juniors Bildarchiv GmbH 5t, Janet Horton 34, 35b, Nicholas Han 21t, Steve Bloom Images 18, 19b, Tierfotoagentur / 10, 51, ZUMA Press, Inc. 53b

Dreamstime
Holly Kuchera 56, Jordan Tan 53, Mikael Males 57

Getty Images
Fuse 57t, Gabrielle Therin-Weise 54, Gary Mayes 12, James Warwick 31, Manoj Shah 6, Paul Oomen 28, Paul Souders 38, Stefan Huwiler 52, Temujin Nana 24

iStock Photos
Abramova_Kseniya 62, Alexia Khruscheva 63, amite 39b, Andyworks 11, chrys35 25b, Dgwildlife 16, GomezDavid 47t, JoeFotoIS 31t, Manfred Rutz 5, Mikhail Sedov 33, polya_olya 36, Svetlana Belkina 33t, USO 17t

Nature Picture Library
Cyril Ruoso 13, Danny Green 25, Klein & Hubert 15, 37, 40, Tim Fitzharris 35

Robert Harding
Frederik 44, Juergen Kosten 45

Shutterstock
schubbel 1, AndreAnita 64, ArCaLu 9, atiger 7t top, Barbora Polivkova 22, Barry Bland 7, Binson Calfort 43b, Brinja Schmidt 42, Chris Balcombe 29, Darcy Ella 46 Eric Gevaert 23, Ermolaev Alexander 4, evenfh 61, Hquality 27, Ian Dyball 60, IanRedding 27t, Jan Mastnik 26, Jeremy Richards 60b, Judita Juknele 14, Kev Gregory 39, Lamberrto 48, livcool 59b, Luniaka Maria 51t, Makarova Viktoria 63t, Mogens Trolle 30, nattanan726 2-3, Photobox.ks 32, Rita_Kochmarjova 37b, 41b, 43, 47, Roger ARPS BPE1 CPAGB 41, Saranga 9b, Shuji Kajiyama 20, Susan Gan 58, Theodore Mattas 19, Turistas 50, Vladimir Turkenich 59, Volodymyr Burdiak 17, WildMedia 11b, Xudong Xia 55, ZiWkun 55b